Subject to V...

Margaret Benson

Alpha Editions

This edition published in 2024

ISBN : 9789364732888

Design and Setting By
Alpha Editions
www.alphaedis.com
Email - info@alphaedis.com

As per information held with us this book is in Public Domain.
This book is a reproduction of an important historical work. Alpha Editions uses the best technology to reproduce historical work in the same manner it was first published to preserve its original nature. Any marks or number seen are left intentionally to preserve its true form.

Contents

I APOLOGIA PRO FELE MEA ....- 1 -

II CLANDESTINE CORRESPONDENCE ....- 11 -

III IN THE BOSOM OF THE FAMILY ....- 15 -

IV CHANCE ACQUAINTANCE ....- 18 -

V THE DESERTED LOVER ....- 23 -

VI JACK ....- 27 -

VII A REGULAR FLIRT ....- 33 -

VIII A FAITHFUL FRIEND ....- 36 -

IX KIDS OF THE GOATS ....- 42 -

X COMMUNITY LIFE ....- 46 -

XI FINISHED SOLOMON ....- 51 -

I
APOLOGIA PRO FELE MEA

WHY were cats created? I do not mean this as a sceptical question, doubtful of any end in their creation; no answer about adaptation and environment would be adequate, nor any statement of specific use. For with all the higher animals—that is to say, with all the animals one intimately knows—there is some beauty of intelligence, physique, or character which renders them, as one must necessarily believe they are, ends in themselves, not only means to the perfection of our very egotistic species. The dog, for instance, has at anyrate moral beauty, and the stag physical; but the cat, who so often loses her physical beauty after the first year of her life, and who slinks about with a weight of strange and secret care on her shoulders, what has she? Who ever knew a cat of really fine character, and yet why otherwise do they suffer such bitter experience? Not experience merely of pans and pots and cat-hunts, which only touch the physical cat; but of the real, keen, emotional suffering of the moral cat, fierce pangs of envy, and the burden of alienated affection? I think cats must be meant to be good rather than beautiful.

When Persis walked out of her travelling-basket, I thought that I had never seen so pretty a kitten. She was about as long as she was high, and as broad as she was long; her coat was of grey—or as this particular shade is called blue—and white, soft, long hair; and she had olive-yellow eyes. She would not have much to say to me just then; but when I came into the room, where she had been shut up in the evening, and saw the little, upright figure sitting on the table beside a lighted candle, which my nurse had set there in case she should feel lonely and unhappy in the dark, after a moment's contemplation—for Persis is shortsighted—she jumped down and rushed to meet me.

She is very well-bred; of course her white is a mistake—she ought to be blue all over; but she has all the other signs of good breeding—long silky tufts in the inside of her paws; ears so beautifully feathered that all other cats' ears look distressingly naked; a little, dark smudge on her pink nose, to show that she knew it ought to have been black; and now she is full grown, the most beautiful tail I have ever seen—"like a squirrel," children say.

She was not called Persis at first, but Hafiz. The popular rendering of that as "Uffiz" was not very pretty; and while the salutation to "the beloved Persis" was being read in the second lesson one Sunday morning, it suddenly struck me that Persis would be a very nice and appropriate name for a Persian cat, and the name "took."

Her manners mostly were charming, and with gracefulness like a well-born lady she would stretch one hand from her basket to greet one coming into the room. She was very affectionate; she would put her arms round my neck in a way I have never known any other cat do, not even her children. Like most other Persian cats, she would kiss me and lick my hand. She had, I will confess, one rude trick: when she was in a larky condition in the twilight, if she caught my eye, she would run, with her head turned round and the side of her face on the ground, all about the room, ending up by coming quite close to me, and jumping and clawing in the air. The position was ludicrous, her head twisted round, and her eyes fixed on mine so that she could not see what was in front of her, and ran sometimes into legs of tables and chairs; her nerves, too, in such a tense condition that if one startled her she would jump high into the air, and then flee into a corner. She always reminded one of the way in which a cockney street-boy makes faces if you catch his eye.

She was not always amiable, the one defect in her character was that she was liable to "strange fits of passion," and would pass from play to anger on occasion without the slightest warning.

She is the fiercest cat towards other animals that I have ever seen. While she was yet a tiny kitten, I brought up a large semi-Persian Tom cat to paint. The tiny kitten chased this big creature round and round the room; if he got under a chair, she got on it, and reached down a little menacing white paw to slap his face. He submitted meekly, until, in order to see what would happen at close quarters, I brought her quite near to him. She spit and swore at him, but thus brought to bay he knocked her over with a sounding box on each ear, and she fled under the table, where, with a tiny drop of blood on her face, she bemoaned herself and appealed for sympathy, the picture of a helpless, injured child. As for the other cat, once roused he went on growling and spitting all morning.

The only small quadruped I ever knew Persis not want to fight was a rabbit. Some children on the place had a tame rabbit which was very fond of cats. One day she met him out of doors. He saw her and came running to play with her; she looked with a horrified face for a moment then turned and fled; she must have thought him a deformed sort of cat; much as if children met a human being with huge pendent ears and an uncouth way of walking who wanted to come and play with them.

Persis was very musical. If one whistled to her she would come from any part of the room, creep up as near to one's face as she could, purr loudly, lick one's face in growing rapture; then, if the whistling continued, she got over-excited, and had to manifest excessive pleasure by biting. I am determined to tell a story which no one will believe, but which is none the less true, that three or four times she has been found standing on the music-stool and making dabs at the keys with her forepaws; she, of course, had discovered before that a piano would make a sound if walked on, and she not unfrequently practised in that manner, but these three or four times I looked up, being surprised at hearing the same note repeated, and found her standing as I have said. However, no one need believe that, and it is their own loss if they do not; and anyhow, now it is a matter of ancient history, for Persis lost all care for the æsthetic part of life when she had a family to bring up.

While she was still an independent lady she used to sleep in my room, chiefly on my bed. It was a difficult matter to arrange at first, because I did not want the kitten to sleep on my face, which was her constant aspiration. Consequently, when I put out the light and settled to sleep, placing her firmly at the end of the bed, a loud purr was heard, and a little dark form proceeded to march up, stamping her paws on the counterpane and drawing them out in rapturous expectation of a pleasant evening.

Finally we compromised: she was allowed to sleep half-way up, embracing my arm if she liked. But I was rather glad when this habit was broken, because she began not to leave me enough room. One of my brothers thought he would try her in his room one night, but he had broken rest; for first she made defiant runs at him from the end of the bed, then in the middle of the night he was waked up by a pitiful howling, of which he took no notice. Two hours later he was waked again by louder howling, and then discovered that the cat had got out of one of his windows, walked on a narrow moulding round to a shut window, and did not dare to go back again. She was so overjoyed at being taken in that she fell into the bath. After that she came on his bed.

But I am wandering from the point of my story. Before Persis' kittens came she had some friends, but no rivals. She treated her friends in a rather severe manner at first. One of them was a fox-terrier, called Don. The first time she was introduced to him she nearly jumped out of her skin with swearing and spitting. When he went out of the room, she went round to all the places where he had been and spit at them afresh. She has a fine scent; if new people have been in the room she always goes round and smells the places where they have been. She smells every new dress I have. The meek Don, who could kill a strange cat as soon as look at it, submitted wonderfully to her whims; and when she flew at him, beside herself with passion because he was enjoying the coffee sugar at the bottom of a cup merely picked the cup up in his teeth and trotted off. But she soon got accustomed to him. And then, distressed at his appearance, tried to lick the black spots off his back; used stealthily to wash the inside of his ears, ready always to rain a shower of blows on his nose with the tips of her paws if he so much as turned round. Then she began to worship in a manner not common to cats; with the sincerest flattery, she used to lie at his feet in the same position that he was lying in; if, for instance, he was lying with his legs stretched straight out below him, she would lie with her back touching the tips of his toes and her legs stretched out in the same way—an unnatural position for a cat.

Now her daughter, the image of Persis, will lie in the same way at Don's feet; but I have never heard of any other cat doing it.

After this she became acquainted with a Gordon setter, and the obstinate curliness of Di's hair gave Persis as much occupation as those black spots on Don's back which never would come off. But she was jealous of none of these, she knew herself to be—as a cat—so infinitely superior to them. She was jealous of nobody and nothing until her kittens came.

There are certain great facts in life which nothing can prepare you for. No amount of reasoning, no previous imagination, will make you in the least able to calculate your feelings. Such must be the moment to very many when they realise that they will die; such is often the moment when people or creatures realise that there exists a little helpless living thing, theirs peculiarly, and yet not themselves. The change that her child can work in a grumbling, egotistical woman is incomprehensible,—could not have been argued by any logic; but far more surprising the event must be to a creature who does not know what is going to happen, cannot guess that her feelings will be moved in a totally new way, and could not realise beforehand that such an event might happen to her as it had to others. I tried to prepare Persis once; I gave her a stuffed kitten on a penwiper to play with. She looked at it with some interest, licked it a little, shook it, and left it; treated it much as a rather careless child treats her doll, but more amiably than she treated other animals. Nor could she dream that little bits of fur,—much like that to the outward eye,—endued with just enough life to tremble on their little weak legs, and utter tiny, plaintive shrieks, should rouse her to such a passionate emotion as to make her forget her own pressing bodily wants.

We know very little more than she did about it, we know just the bare fact that it always will be so, but why it *should* be so we know no more than she. Who understands the miracle by which an utterly selfish creature, whose natural instinct is to hate all other animals, and, indeed, only to tolerate human beings because it can make use of them, should be made to know and feel, in a short ten minutes' space perhaps, an overpowering, passionate, protective love?

One morning Persis did not feel very well, in sign whereof she showed a decided intention to occupy my bed. She was sent down to an empty bedroom while a hamper of hay was being prepared for her; but when her invalid couch was ready she was nowhere to be found; a search discovered finally that she had put herself to bed in the room already, under the counterpane. Still, she was thinking of nobody but herself. Later in the morning I visited her,—when three little helpless, shapeless, furry things were moving about her, and Persis was not thinking of herself at all. One would not have believed an animal's expression could change so much; the overwhelming surprise, the intense affection, were in her face as clearly as they could be in an human face; for the time her egotism had gone, she was not a cat, she was a mother. Formerly she had been shy of people, frightened of men; now, as one after another came in to see her kittens, she showed no fear, and, what was even more curious, no anger; she merely purred in pride and entire confidence.

They were wonderful kittens—two quite blue, one like its mother; their eyes were shut, their ears were flattened down over their faces,—they were little bodies which breathed and fed and grew.

But they *did* grow, and their ears stood up and their eyes opened,—dark and light blue,—and their heads got steadier, and in a month they were little square solid kittens, who with much difficulty could get out of the box in which they were placed. Getting out was a process which involved the fullest exercise of all mental and physical powers; for first they had to advance to the side, then one tiny paw and then another was put over the side, and the adventurer was for the time hung up by his shoulders. Then he worked himself on by the help of much kicking behind and clawing against the box, until the part outside was just heavier than the part inside, and with a scramble, and by the help of the centre of gravity, the whole kitten tumbled on to the floor. It was a grand triumph of mind over matter. And still Persis beamed on them, and on the world in general.

But as they grew began the first little rift within the lute. It was difficult to help it. I put it to you—could one carry three kittens and a cat about, like Henry III. of France, to exhibit to visitors. If it was a choice between exhibiting kittens and cat, visitors would surely prefer to see the kittens; and so it came to pass that the children were carried into the drawing-room and handed round, while in the empty schoolroom the "old" cat sat alone. It was only a couple of months since she had been shown to visitors herself. Sometimes I took her too, but that was not a great success, for everybody liked the kittens best.

And now the kittens began to be steady on their legs, and able to run and play, and their horizon was no longer bounded by licking and feeding and warming; and when they once began to play, their mother seemed rather large and rather old to play with them. Persis did not care to play with me or cheek me any more, but she liked to gambol with the kittens. So she played mouse in front of Pasht, but Pasht would rather play with her brother and ran off the other way; and she pretended to be a tiger lying in ambush to wait for Marjara, but Marjara wished to tie herself up in a soft heap with Ganem and bite his ears, so the Old Cat stopped in her gambols and looked at them.

Ganem was given away; and as he had been rather a favourite playfellow, and the least favourite child of his mother, the family got on more happily after that. Then I went away, and saw them no more for some two months. When I came back, the Old Cat and Pasht were sent for.

They made their journey in a large hamper, and were brought up to my room. Pasht had grown lovely; soft mouse colour with topaz eyes; but nevertheless the meeting was a real disappointment. Persis came out of the basket and with no greeting to me, jumped down and went to look out of the window. What could I do? I had to play with Pasht.

I thought perhaps the cat's temper had been upset by the journey, so I left her alone, and some hours after came back to both of them. Persis was lying and staring out of the window, and the kitten was occupying the room; it ran at me, jumped and climbed up with loud purrs, and rubbed against my face. I went to the window-sill, and still Persis did not move; when she saw the kitten she growled a little; I put it down close to her, on which she spit, slapped it, and fled.

So things went on. When I came into the room the kitten always ran to greet me: it was impossible to take no notice of such a soft, confiding, mouse-coloured creature, yet all the while I was speaking to it two great sullen, green eyes were fixed on us, watching us round the room. If I came there to speak to the cat, she went quickly away, if the kitten approached her she spit, and if it came nearer, hit out at it. Evidently the change had come in Persis from a kitten to a cat. She was a mere domestic cat, with a not very amiable temper, she cared no more for human beings, and had arrived at the queer alienation from the young when they are grown up which comes to nearly all creatures; she had had half a human soul once, but she had fulfilled the animal functions, and she was an animal again.

Yet one or two symptoms seemed to belie this view. Once or twice, coming into the room, I greeted her first. Then she purred until the kitten came near, when she got up and left us hastily.

But it was difficult to see why this sullenness should so perpetually prevail. She hardly ever forgot it. Her big green eyes had almost always that sullen, lowering, miserable expression.

Now and then, indeed, when twilight came on, she rushed in and out of the room, alternately defying the kitten and flying from it; but not the most unimaginative cat on earth can resist the excitement of the growing darkness, when the eyes flash out in amethyst and topaz, and the pupils dilate with dramatic terror and eagerness. But twilight deepened into dark, and candles were lighted and fairyland stopped, and the legs of the tables and chairs ceased to be tree-trunks in a jungle, and Persis came back to life in the schoolroom, and despair clouded back on to her brow.

But the truth only began to dawn upon me one day. I took Persis into my own room quite alone, and suddenly the sullen expression vanished; I carried her in my arms and she began to purr; I put her down and she walked up and down on the counterpane, stamping her paws and spreading her claws,—Persis had all at once become a kitten again. She licked my face and put two arms round my neck when I took her up. I brought her downstairs, thinking our old relations were re-established; the kitten came near, and Persis walked hastily away from me and took no more notice of either of us.

Then the kitten ailed and was sent away to be nursed, and with that curious, confused idea that creatures have, the mother felt a lack somewhere when the object of such strong emotion was removed, even though the emotion was only jealousy. She hunted for the kitten all afternoon. We found her in a part of the garden which she did not usually frequent, and she ran away with a sense of guilt when she saw us. But when evening came, and she was in the room alone with me and there was no kitten, I was left in doubt no more as to what it was which was moving her. She squeezed herself in by me on the sofa, she kissed me and purred blissfully.

And so it goes on. I have not had the heart to banish the kitten altogether, yet when she is there I can seldom get a purr or a look from the cat. One day I persuaded her to let me stroke her under the ears and the throat; this is almost like mesmerism to a cat, and if one can persuade them to let one begin, one can do almost anything with them; and so I was gradually bringing her to a happier state of mind, when the friendly kitten, perceiving that something sociable was going on, came up to share in it. They met face to face as Persis took turns up and down under my hand. They looked at each other for a moment, then she slapped the kitten in the face and fled.

What am I to do? If I keep the kitten I cannot prevent this jealousy. Persis lives in a condition of perpetual, jealous misery; if she thinks the kitten is sent away, or that she is exclusively favoured, then only does she emerge out of sullenness. And yet she is not really devoted to me; she is only a complete egoist, and cannot be happy unless I am devoted to her. After all, am I not bound to her? Was she not once my sole and only cat, carried about, exhibited to company, hunted for if she got lost? And yet Pasht is much fonder of me than Persis ever was; Pasht will run after me, while Persis wishes to run away and be fetched back. Pasht comes to meet me when I come into the room, cries to be picked up, purrs as soon as I touch her; but when I do so, those green, miserable eyes watch me, and Persis will allow no caress which is not offered to her first.

What shall I do?

II
CLANDESTINE CORRESPONDENCE

THE last week has been an arduous one; I have had to chaperon Pasht.

Pasht has experienced her first proposal. I suppose it is no wonder, considering her age, that she was flattered; but I could wish that she had fixed her affections on anyone less vulgar and under-bred.

This was how I found it out. Pasht had been for many days very eager to go into the garden. One morning we were playing croquet on the lawn, and I paid no attention to the kitten, until suddenly I looked up to see her lying on the path, her long thick hair fluffed out, her sweet mouse-coloured cat's visage resting on the edge of the grass, her little chin rubbing against it, and her long squirrel tail lazily sweeping and thumping the gravel.

At first I thought it was only flirtatiousness in general, an attempt to captivate the universe at large, when lo! out of the laurels opposite to her flashed an ordinary, vulgar, ill-bred, short-haired tabby cat, who stood there for a moment, looked at me and disappeared.

I was very much shocked, picked up Pasht and shut her up in the schoolroom, when she instantly appeared on the window-sill and reproached me loudly. But of course I did not take it seriously, and thought that they would both get over it.

I must explain the position (unfortunate in this respect) of the rooms in which the cats and I live.

It has four large windows looking on the lawn and the laurel bushes—too high for a cat to jump down, but not too high for her to practise little wiles on the window-sill for the benefit of appreciative spectators below. Just on the left hand of the door is a long window, from which steps go down to the garden, and close by the steps is a large laurustinus, a most convenient place for ambushes and clandestine meetings. Opposite the schoolroom door, again, there is another door opening on to a back staircase, whence one gets into kitchens, whose windows also give on to the lawn, and are usually open. My bedroom is above the schoolroom.

On the evening when I had abruptly stopped Pasht's flirtation, a noise arrested my attention as I was going to bed. It was the voice of a cat saying "wwoww." You know what it means when a cat says that? He is paying compliments. The noise went on and on, round the schoolroom-end of the house, until I went to sleep, but I heard no answer from Pasht.

Pasht was hysterically affectionate when I saw her next morning; she said "a - - - ow," and clung on to my dress, and climbed up on to my shoulder and refused to leave me, and walked about over my letters when the ink was wet, and flapped her tail into my mouth, and altogether played the fool, and pretended that she had forgotten her vulgar suitor of the night before and I heard no serenades outside.

But in the middle of the day I suddenly heard from my bedroom an extremely loud voice saying "wwaughwow," and looking down saw Pasht standing on the window-sill of the schoolroom. I don't know whether she said it or not, for as soon as she saw me she looked up and took to the more ordinary and ladylike expression of a general desire to go out in the sunshine. Several times in the day I heard it again, but as soon as I looked round, Pasht turned an innocent face to me and said "miaow."

In the evening the gentleman began to woo again; I knew it was the suitor this time, as Pasht was safely shut up. I listened at the door of the schoolroom to hear if she was answering, but there was no sound. She *is* a regular flirt.

A party from the house went round the garden with croquet mallets, but with no result.

Next morning it became too clear that Pasht was encouraging her suitor; he rushed away from the laurestinus bush as I came out, and she was sitting on the window-sill. I took her out for a short time in the garden under strict supervision, but she would do nothing but flop into graceful attitudes on the lawn. I really had not thought it of her.

I took her in again, and argued the point a little.

I told her that she was behaving in a very vulgar and forward manner, and that no nice Tom would respect her. She merely looked up in my face and said "a - - - ow."

Then I said I would not have made any objection if he had been a gentleman, but he was so exceedingly common and ill-bred.

But she still looked with pathetic topaz eyes, and opened a little pink mouth with a deprecating mew.

I felt much as if, "with a little hoard of maxims," I was "preaching down a daughter's heart."

And what was worse, it did no good. Every time the door was opened, however much Pasht was pretending to be devoted to me, she suddenly found she had urgent business in the kitchen, and flew downstairs; and when I, knowing the nature of the little flirt, did not go down to the kitchen at all, but straight out of the long window on to the lawn and found her there, she looked up with the most innocent face possible,—"Yes; after all, I see you enjoy the sunshine as much as I do." When, in spite of kicks and struggles, I carried her in, she never once said "wwoww," but merely gave vent to the emphatic mew which means, "*I don't want to go in.*"

I took her an airing in my arms that day, but it was extremely exhausting, and I covered my dress with long hair.

And all that night the cat mewed.

Another exploring party went from the house with shovel and tongs.

I couldn't stand it any longer. Pasht was sent away to a very strict boarding-school system at the farm.

A week after, when the strange cat had ceased to howl round the house, she came back again; but as soon as the schoolroom door was left ajar, the urgent business in the kitchen claimed her, and Pasht disappeared for many hours.

Poor little Pasht, were you disappointed that no one met you in the garden to flirt with, or wanted to bounce out of the laurel bushes and exhibit his masculine beauty before you? Or, after all, is your little heart as hard as I think it, and do you prefer a nice warm room, a lawn to romp on, someone in whose lap to lie, who will gently ruffle your throat and ears—do you really, deep down in your heart, prefer these beyond all lovers whatever?

Anyhow, when Pasht appeared at the long window, she had a gay, innocent little air on, and she ran in saying, "You see, the fine weather *did* tempt me to stay out rather long,—where is my breakfast?"

Never mind, little Pasht; we will arrange an honourable alliance some day with a gentleman of rank.

III
IN THE BOSOM OF THE FAMILY

IS it not true that there is a very general want of recognition of family-life among domestic animals? It is a great mistake to suppose they are incapable of it; often, as a matter of fact, they do not lead domestic lives, for the simple reason that people will not let them. If, for instance, you won't keep a whole family of cats, how can you expect them to develop domestic affections? We talk of their being "domesticated," but we mean that they are made a part of our domestic arrangements, without being allowed to have any of their own; yet they are quite as capable of it as we are. Of course their domesticity does not last long, naturally and necessarily not, because they have not one family but a series of families, and one family must be dismissed before the next is taken on; so domestic affection develops into murderous desires. However, I must say that in all experiences I have personally had of cats, guinea-pigs, rabbits, dogs, goats, and birds, I have only known one murder, and that was by an uncle.

Rector was allowed to have all his family about him. His wife was decidedly under-bred. He was called Rector, in fact, because he would not catch the mice, and had to have another less aristocratic but more useful cat to help him. The curate was called Jenny. She was a low-bred tabby. Rector could not help despising Jenny, and if anything vexed him he used to bite her badly; but she was a very meek drudge, and took it as a matter of course. Rector was white, with blue eyes, so we only kept the white kittens, some of which were blue-eyed, and *not* deaf; blue-eyed or not, Rector used to take them out walks in the evening.

The four—papa and mamma and two kittens—used to proceed together to the mound near the pump, and Jenny then left them, to crouch in the bushes,—this for a purpose of her own.

Then began the game. Rector rolled the kittens over and played with them gently, until all three became a little excited; then, if Rector got carried away, and bit or scratched his infant till it squeaked, out bounced Jenny from the bushes to deal him a handsome box on the ear; and, having thus admonished her husband to take better care of the children, she retreated again to the shelter of the yew-trees.

If you keep a whole family, you will find that there is not only a parental, filial, brotherly, and sisterly relation, but also a grand-parental. When Midge had some white kittens, Jenny, whose under-bred offspring had been put out of the way shortly before, helped her to nurse them, with as much pride and perhaps more solicitude than Midge herself showed. It was a most charming scene. We went to see the family soon after the birth of the kittens, and found Midge, in the rôle of the interesting young mother, leaning back upon Jenny. Jenny put a paw round her, while they surveyed— the mother languidly and the grandmother proudly—the squirming white family.

But it is not cats only who have these strong domestic ties; almost every animal shows the same thing in a greater or less degree.

We inherited, on changing our home, a beautiful pair of swans. The first year that they became ours they had four cygnets, and brought them up extremely well. It is true that when they were full grown, the cock-swan, if one may use such an expression, tried to kill them; but that was only natural, they had become his rivals. They were variously disposed of: one was taken up to a pond in London, from which, not being properly pinioned, he escaped, and kept a cockney crowd for an hour well amused on London Bridge by flying over it and swimming under, after which he— or, as he could not possibly be caught, the abstract idea of him—was presented to the Thames Conservancy.

So far, this doesn't seem to have much to do with the swan's idea of home, but, as some candid preacher said, "You may think this has not got much to do with my text, but I'm coming to it presently."

The swans lived on in peace and happiness through the autumn and winter, but in the spring, when they ought to have been nesting, some wicked boy hit the lady swan on the head with a stick, and she sickened and died.

For some time the widower was left solitary; then we thought this was rather cruel, and busied ourselves in getting a mate for him; and a fine young swan was procured. When lo! it was found that the old fellow would not let his young companion come into the pond. We thought it would "wear off," and left the young one to its fate; and many times we passed the pond to find the poor young thing squatting sadly on the road, and the widower swelling up and down.

Then we found there was a slight mistake, the young swan was a cock-swan also.

So we changed him, and got a real lady instead. This time he would just let his companion come into the pond, but oh! she had a bad time of it there; he pulled her feathers out, and he drove her away from the bread; but it had to be gone through,—it was his way of showing constancy, and it turned out all right. She is treated now with as much respect as his first wife.

But she was a very young wife; so, when she had hatched three eggs into cygnets, her pride knew no bounds. The father, getting into his dotage, encouraged her in her maternal follies. The cygnets were fine healthy birds, but the two old birds took them out walking to such an extent that one by one they died. No one quite knows why. Some say that there was not enough grass by the pond, and the parents took them to find grass; and some say that parental vanity wished to display such flourishing offspring; but anyhow, the fact remains that the cygnets took walks with their parents till they died. There is nothing more domestic than the family walk.

But now contrast this domestic affection with the melancholy fate of the inebriate swan.

A clergyman's wife kept one swan, and the swan, no one knows how, got into the habit of going to eat malt at a public-house. If he had done this within bounds it would not have mattered, but he got regularly intoxicated, and every evening reeled homewards. His mistress tried to reform him, but to no purpose; and she tried to shut him up, but he got out; and she used to meet him coming home with rolling, uncertain step and hanging head. She wept, for it was such a bad example to the parish; but that had no effect on him. At last, one evening, he was run over and killed while reeling home in a state of intoxication.

Now, how far more melancholy is such an end than that of the three infants killed by family affection! I would rather die three times over from walking with my family than once from intoxication.

What is the moral? Do not break up the family too early. The presence of the children (up to the age when he wants to kill them) will have a softening and steadying effect on the manners of the father; while who knows what stores of masculine experience he may not impart to his children up to the time when they wish to fight him.

Besides all this, it is really much more amusing.

IV
CHANCE ACQUAINTANCE

HOW vividly one sometimes retains for years the memory of a chance acquaintance—a person whom one has met but once, passed in the street, talked to for half an hour, whose name one may not even know.

A friend of mine was travelling in Persia, and as she and her brother were resting in a caravansarai after a journey, they saw a Persian gentleman beckoning to them from the garden. They went down to him, and he asked them to come and have supper with him. They came, and found the bread laid out, plate-wise, and the roast meat on it. They ate and talked to him, and after their meal went on their journey. They never asked for nor heard his name, nor he theirs,—they will never meet again; but that Persian gentleman will be as vivid to them until the day of their death as a friend of years.

Such memory of a mere passing chance acquaintance is not confined to human beings. Sometimes one meets animals for an hour or two, sometimes one accidentally lights upon them in a crisis of their lives,—such even as their death,—and one suddenly and unexpectedly understands and knows them. Some people and animals one never gets near. You may, for instance, sit opposite people in church for years, know all their Sunday dresses and hats, and how much they give in the offertory, and be not a bit nearer to them in the end than at the beginning. Such is the acquaintance one has with caterpillars; they are always just the same; they eat and grow and become cocoons, and reappear as butterflies, and there is no character from beginning to end. That is partly why they are such excellent symbols.

Then there are some animals that have no sense of intimacy; they let you into all their domestic relations,—their committees, their politics, and so forth, at once; for the reason that they have only one side to their character. They have established a Platonic Republic; they do their domestic duties on the scale of the commonwealth, have a universal nursery and government education. In spite of their monarchical arrangements, they are real socialists at heart,—they care for nothing but the good of the State. Even those that live in a tiny community, two or three together, have no real individuality. Have you ever found one of those tiny round nests, like ashes of paper, which apparently grow on a stalk, and in which two or three yellow and black tree-wasps live? It is the easiest thing in the world to scrape acquaintance with those wasps; kill an ordinary housefly and give it to them. They will take it from your fingers, and, without the slightest shame at "talking shop" in public, will roll it into a neat, hard, black ball, crushing up legs and wings alike, and stow it away inside the nest.

But the want of intimacy characteristic of many insects is not characteristic of insects *as such*. I once attended a grasshopper crisis. There was nothing professional about the grasshoppers; they did not not "spend themselves in leaps ... to reach the sun." They did not think the least bit in the world about the sun, they were merely private individuals—courting. Grasshoppers' courting is an organised affair. I saw it in Switzerland on a soft, sunny afternoon, when the hotel population was divided between the Roman Catholic Church on the right and the English Church on the left, and the steps of the hotel between the two. As I dawdled along by a bank of bilberry just turning red, the grasshoppers were singing loud among the stalks of heather; suddenly I was aware that they were not singing aimlessly and jumping without purpose, but that they were intently engaged. It was like the old fairy-story, when a child falls asleep on a bank, and wakes to find himself surrounded by fairies intent on preparation for the marriage of the king. The large limp ladies were sprawling about ungracefully, and in front sat their small, spry gentlemen singing away. Here was a green gentleman serenading a brown lady, and I wondered at his taste; presently she got up and ran away. Clearly that was part of the drama; it was the genuine "flirtatious" instinct of avoiding a plain answer on purpose to provoke pursuit; for the gentleman does not jump, but runs after her to bring her back. When lo! a green lady is seen crossing the path, also coyly escaping from a suitor, and the faithless swain is captivated all in a moment by the green charms, and deserts his brunette to pursue her. Further on— astonishing sight!—is a young ladies' school, just "come out"; fourteen or fifteen green and brown ladies, shy and awkward, scrambling down the bank and all talking together.

I never saw such courting before or since, but I shall never lose the feeling of intimacy, for I know now that grasshoppers are not always little machines arranged with the greatest amount of muscle for the smallest amount of weight, or wound up to trill on in the sunshine, as mechanically as a watch ticks, or even created to be a burden,—but they are tiny creatures, full of emotion and insect loves, putting their best energy into their whirring song to claim the admiration of the languid, lovely creatures that lie lazily listening.

But sometimes one arrives at a sudden personal relation to a wild creature, too often ended abruptly by its escape or death, and its kinsfolk are never afterwards to one as little as before. One has regarded it as a member of a class; henceforward one regards that class as composed of individuals possessed of strong personal desires, needs, emotions, not merely obeying what we call "instincts,"—meaning thereby the mechanical impulse to eat, grub, make nests, care for young. To take an extreme instance, perhaps you think that moles are altogether uninteresting, merely existing for the sake of lightening the soil and destroying the wire-worm, and, in case of undue increase, fit to make a cap for the mole-catcher and a little skeleton to swing from a tree. But perhaps some day you will see in the stubble, after the hay is cut, a little black form running confusedly round and round; catch it, and hold in your hands the soft, velvet-coated body; feel the funny, groping snout pushing through your fingers, on the chance—however different their touch is from the damp, delicious earth—that it will be able to find some place where it may grub a hole and escape; realise that you might make a pet of this small, soft thing, and then please recognise its wild desire for liberty, and let it go.

But there are some animals which, although usually recognised as "wild animals," seem to have no fear whatever, except when they are being chased; once they are in the hands of a human being they are completely self-possessed. A friend of mine sat in a field when the hay was being carried, and saw a little field-mouse playing about; she pursued and nearly caught it, but it finally escaped. She came back to where she had been sitting to fetch her umbrella, and under it was found another little field-mouse asleep, which she caught without difficulty, carried back, and put into a box with holes in it.

She brought him in to tea that afternoon, and even at this, his first meal, he sat up like a kangaroo on his long hind-legs, and ate bread and milk out of a spoon. He absorbed alarming quantities of it, fell instantly asleep, woke up after a few minutes and ate a great deal more; but the next morning the poor little beast was found gasping, apparently dying; and when his box was opened he would not run away. But he presently recovered as suddenly, and again devoured much food, and so went on through the day, though his gasping fits returned at intervals. Next morning he died. Is it that we find these creatures generally when they are ill?—the least touch seems to make them die. Certainly I remember once or twice, in those joyful days when sitting in a hayfield meant the height of bliss, that our very gentle and amiable collie, excited by an "animal" smell, would grub open a nest of little field-mice, and stand by delighted and smiling at his discovery, while we came up just in time to see three or four expiring infants. He could hardly have killed them, for he only wanted to look at them. Yet they died.

What was it, I wonder, that killed Maximilianus? Maximilianus was a very small shrew, and we found him running about the garden; he was just about as long as his name. He was not the least frightened, and we carried him about for half a day; but we found nothing he could eat, until at last we came upon a very large, fat, orange-coloured centipede. Maximilianus seized upon this with the utmost delight, began it vigorously at one end, and ate it up like a radish as far as the middle. Then he died.

We had once a visitor in the shape of a squirrel, who came uninvited, made his abode with us for some months, and finally departed, taking "French leave." My mother was his guide, philosopher, and friend. He slept in a pocket of her apron (this was in the seventies), whence he came out to fly up the curtains and drop down, venture on to the breakfast table, and experiment on her tea with a tiny paw. He always ran up the curtain when he was scolded; as for instance when my father, going to the sideboard to cut ham, found the squirrel's head just coming out of it, having eaten its way through from the other side. Then, after being received in the bosom of our family, after sharing meals with the household, after attending lessons and even prayers (when he ran up the back of a kneeling housemaid), the skwug suddenly disappeared without warning. A few days after, my mother was walking in the wood, when a squirrel ran up to her, put its paw upon her foot, looked her in the face, then turned and ran away. It was never heard of again.

Sometimes you find animals which, though not very near and dear to human beings, have a great influence on other animals. Our donkey died the other day. She was a remarkable and original animal. Though she was a fixture, taken at a high valuation from our predecessors, her demeanour was such that we called her Jack, and thought she had retired to a well-earned repose. Then we found she was not quite two years old, and a lady. We were always good friends, but not specially intimate. She and her mule-foal might come to the window for bread and salt when the horses were not allowed on the grass; but for weeks together she did not avail herself of this privilege, till one day a snort was heard from outside, and the donkey's nose was seen flattened against the glass. Once, when my mother was walking with a friend of hers,—not an acquaintance of the donkey,—Jack, for I cannot help calling her so, solemnly accompanied them all the afternoon, walking between them. But such occasional walks, and the fact that she was amiably willing to follow anyone quite impartially for a handful of oats, constituted the extent of our intimacy with her. Not such was her relation to the other animals. As exclusively as my goat walked with the cows, Jack walked with the horses. She did not, of course, consider herself so superior to her company as the goat. She made many friends among the horses; you might not have known it, perhaps, but neither as a general rule would you suspect the friendship which men have for one another by their way of behaving. If a man meets a great friend in company, he either takes no notice of him or stands near him without saying anything. Jack used to stand about with the horses without saying anything, but they liked to have her near.

One morning Jack was found dead of fatty degeneration of the heart. "I'm sure the horses miss her," said the bailiff's wife; "I look at them standing in the yard, and I can see they miss her."

Jack was buried in the orchard, and her little mule followed the body as far as the garden-gate. But there they shut the door, and the one mourner was left outside.

V
THE DESERTED LOVER

EVER since I was a very small child I had longed to possess a pair of budgerrygars. There was a tradition of three live ones once in our family, in proof whereof my nurse could point to a little stuffed bird in its case. I used to gaze with longing at that beautiful green and yellow creature, with the speckled back and the black and blue feathers in its neck, sitting with a foreground of quaking grasses and an eternal blue sky behind. There existed also, but rarely seen, a little cardboard box containing a few of these same mysteriously beautiful blue and yellow and green feathers, with here and there a long strong tail or wing quill. Yes, there had been budgerrygars among us once; there were even real live ones now in the possession of those happy Italian women who sit at the street corners, but for me—while I was still a child—they were inmates of that imaginary Paradise of unattainable things, wherein might be found little wax cages of birds, and the fluffy hollow ducks which live in confectioners' shops and are sold for ninepence.

After I was grown up, a friend gave me one of these ducks; I have it still, and the halo still surrounds it. When I was grown up, too, some one gave me a pair of budgerrygars; and there followed a tragedy which was not bargained for in the price paid.

They came down from London in a tiny cage,—a travelling cage. Budgerrygars do not mind lack of room, it makes it all the easier for them to sit quite close, as if they were glued together. They were lovely little things, with their pearl-grey beaks,—wonderfully sharp and strong those beaks are, as I know to my cost,—but they could use them gently, and you would see one turn with a soft croon to put straight a ruffled feather on its mate's head.

The little gentleman had caught a cold, not much of a cold at first; he only panted slightly as he sat near the little lady and ruffled his feathers; but she cheered him up, and smoothed the feathers down, and they sat side by side and looked at the world with little meaningless grey eyes.

Their new large cage was a great excitement, and it was immense fun for them to walk over the top, using their beak as a third leg, and that the most reliable. And their spirits ran so high that they began to shriek unmusically at each other when they found themselves at opposite corners of the cage.

I am afraid we were not as careful as we ought to have been with the little gentleman. They were so funny and pretty that they were carried from room to room; and the cage must have been in a draught, for the little gentleman began to puff and breathe rather hard, and his feathers were persistently ruffled, and the little lady could not smooth them down any more, even if she had tried.

Sympathy to the ailing, the feeble, and the weak is a very modern virtue; strange, as civilisation shows us what an unprogressive virtue it is. The lame and the blind were "hated of David's soul"; animals and savages and men of early civilisation agree with David. Now and then you find a dog which will bring a broken-legged friend to the hospital, a cat which brings its half-starved neighbour to eat its own dinner,—souls of philanthropists on pilgrimage, dead or yet to be; but the stag's instinct of goring the sickly ones, and the wolf's of tearing the wounded, are the ruling instincts. The lady budgerrygar took David's side in the matter. She did not wish to bite her spouse, or peck him, or pull his feathers out, but he began to be hated of her soul.

One day she would not let him sit by her on the perch; he could hardly get up to it, yet he would have done so for the sake of sitting close to her, for the sake of putting a stray feather straight in her ladyship's top-knot, of feeling the little pearl-grey bill travelling softly over his head with a croon of affection; but she would not have it, she drove him away from her. So he sat on the lower perch, or on the bottom of the cage; he did not scream or croon, he just puffed his feathers out and panted. Did David repent in respect of the blind and lame when he said, "My lovers and friends hast thou put away from me"?

What strange rebellion against fate moved in the soul of the little budgerygar, what necessity of finding a lonely place to die in, what sad desire of escaping from the mate who would no longer care for him? It is all very well to talk of "instinct" and dismiss the case, but how do you suppose the abstract idea of loneliness in death nerved the failing wings and feet to seek the door of the cage, made him squeeze through the door, such a little way open; how did it attract him across the room and through the half-open door,—away—away—as far as he could go from his faithless love? Did this abstract idea act on the little budgerygar like a machine, and move and nerve the wings for such a flight? Or was there distress in the heart, and anguish in the little animal soul, when he found himself ill at ease and ailing, deserted and repulsed?

It is a work of skill and time to induce a healthy budgerygar to leave its cage; but this quixotic spirit found his way out of the cage for himself, and found his way out of the room, and he must have flown until he dropped dead. For we found a little heap of gay green and yellow feathers in the passage,—stone-cold and stiff;—he had been dead some hours.

Budgerrygars are very sociable birds, they cannot live alone. The little dead bird could not. So we got a new mate for the lady, whom she received warmly, and the pair lived quite happily ever after.

But I should like to know, in the whole scheme of things, what is the recompense for the little deserted lover.

VI
JACK

FEW people know how different one bird is from another of the same kind. Of course we can see when one canary is green and one yellow and one crested; but few people know that some canaries have blue eyes, some brown, and some grey; or how different one canary is in intelligence and character from another.

Jack was a remarkably intelligent canary; one always felt him to be immensely superior to oneself. When he consented to sit on his swing and allow me to swing him, he always seemed to say, "This is a very childish game, but it appears to amuse you, and I am by nature indulgent." He was often very angry with me and pecked me, but I was sure I deserved it. The only blemish I ever found in him was that he was rather unscrupulous and ill-tempered, but then he was so exceedingly superior that he had to find fault with the canaries and me sometimes.

Jack was very bright yellow, with a slim, trim figure. When he was about two years old a little wife was given to him. She was almost white, and they looked very pretty together. Her name was Thyrsis. We tried to call them Corydon and Thyrsis, but "Jack" suited him so well that we were not able to change it, so they remained rather inharmoniously "Jack and Thyrsis" to the end of their lives.

I always used to turn Jack and Thyrsis out of their cage when I was cleaning it. One morning I did not see that the window of the room opposite was open. They flew round the room together, then coming to the open door they darted out of it, into the next room and straight to the window. One instant they rested on the window-sill, then like a flash of sunlight and moonlight they were out into the sunny garden and trees beyond. All that day I haunted the garden, too anxious to cry, carrying their cages about, in the vain hope that they might be hungry or thirsty and want to come back; once I thought I saw a flash of gold, but night fell and still the birds were out. The next day we sent the town-crier round shouting out a reward of five shillings for them, and the day following Thyrsis was brought back to me in a paper bag, much exhausted but not materially worse.

I did not hear of Jack for five months.

Then a boy who lived near and kept canaries heard for the first time of my loss, and he sent me a canary which some months ago had come through the open window and settled on his own bird's cage. Of course it was Jack. He had not forgotten his way of coming towards me with wings outspread, uttering the funny scolding noise from which he got his name.

Now by this time Jack and Thyrsis were come to years of discretion, and it was thought that they ought to build and have young. So they were provided plentifully with horsehair and cottonwool, and given a small round basket in one of the cages, and we put their two cages together, opening the door between.

They were very much delighted with the wool, and played with it a great deal, but they seemed to have no idea of the proper use of it; if we put it into the nest for them, they merely pulled it out again.

This became so hopeless, and I was so anxious to try to rear little canaries, that a friend promised me another hen. She, however, forgot what our circumstances were, and sent us a pair, who were promptly named Jock and Mummy. I would not have Jack defrauded of his wife after all, so Mummy was taken away from Jock and given to Jack instead. There is not much to tell about poor Jock. He was a middle-aged gentleman, subject to chronic asthma, and could never in that state of health have undertaken the cares and responsibility of a young family. His cage was always hung up near the fire, and when he was worse than usual I gave him a tiny drop of sal-volatile in his water. He was a contented, cheerful bird, and lived as long as with his age and asthma one could expect.

Mummy was a crested bird, pale yellow with a green crest, rather pretty, but in mind utterly vulgar. Of course she was far more effective than the refined Thyrsis had ever been. She knew all about nest building, and began at once; while the cynical and gentlemanly Jack looked on. The pair always reminded one of an aristocratic philosopher who had married his cook.

But one must give Mummy the whole credit of the nest; she put the moss and hair and wool into it, she squatted herself down in it, turned round, fluffed herself out to make it hard and round and compact; and at intervals went to keep up her strength by taking her "dishing-up beer" in the shape of hempseed.

Then she laid eggs quite satisfactorily, and they came out quite satisfactorily, and one by one all the nestlings died—*not* satisfactorily. On examining the little corpses, we found that they had died of starvation. Jack was found guilty at the inquest, for a first principle of domestic life among canaries is that the father feeds the birds while they are very young. What was the reason, then, that he had so disgracefully neglected his duty of feeding them, while his devoted wife sat on the nest to keep them warm?

There must be something more than grandeur and cynicism to make a gentleman allow his children to die of starvation.

At last we found out the reason—Jack was flirting with his first love! Thyrsis' cage was hung in Jack's sight, and instead of feeding his infant children, or attending to them in any way, he clung to the corner of his cage all day and serenaded Thyrsis. We put Thyrsis out of his sight; Mummy laid a second set of eggs, and Jack attended to them as if he had done it all his life. It is true that he threw the eldest out of the nest on to the floor of the cage, but there is great excuse for that; a gentleman of refined and fastidious feelings must have had a dreadful shock when he first saw an unfledged canary and realised that that repulsive creature was his progeny. With all his cynicism, he could never have imagined that anything so loathsome existed. I don't see what else he could have done,—I should have done it myself in his place. From whatever point you look at them, unfledged canaries are altogether and absolutely hideous; their brownish-pink skin is scantily covered with hairs, little bits of flesh wave helplessly about where their wings and legs are going to be, they have two large dark swellings where their eyes are going to be, and the only thing that is defined about them is a huge mouth which is almost always open and yelling. I had to pick the canary up from the bottom of the cage, and I still owe Jack a grudge for it, though I cannot in justice blame him.

Little canaries, when they are fledged, are as pretty as before they are frightful. These three little birds, when they were fledged, were all different and all beautiful. One was like her mother, yellow and green and crested; one like his father, all yellow; and one a sort of mixture, green and yellow and without a crest. Now a curious thing happened: the father chiefly devoted himself to feeding the little hen, who was like her mother; the mother (who begins to feed the birds when they are getting fledged and do not need warmth so much) fed the little cock like the father; and I have sometimes seen these two of their superfluity feeding their neglected brother. He throve well on the little attention he got.

I brought up several nests-ful. We had Tweedledum and Tweedledee,—Tweedledee's name was subsequently changed to "Jewel" by a little cousin to whom I gave it, and who considered it a priceless treasure,—and Daffodil, the neglected nondescript, and Vicary, and Roumenik, called after the Wallachian country-place of some friends of ours; and others whose names I forget. Roumenik was the only one I kept, he was the last hatched, and was called "the Baby" until he died at the mature age of eight years.

There was one wonderful chicken who did not live to have a name. He was very precocious, and died young. This was how it happened: the misguided Mummy laid an egg in January, and in consequence, as I have always believed, of the weather being so much too cold when it was hatched, the bird could never get fledged; when it had already begun to be active and of a roving disposition, it still had no feathers on. Even sprouting wing-feathers might have broken its fall a little, on the many occasions when it tried to get out of the nest and fell on its back on the bottom of the cage. One day it had a fall more serious than usual, and till evening it sat on the edge of its waterglass with its head hanging down and its neck apparently dislocated. In the morning I found it dead in the waterglass. So I do not know to this day which accident it died of.

But meanwhile a sudden stirring of domestic instincts came to Thyrsis, and she was stimulated to rival Mummy's nest-building. I gave her a little basket and materials for a nest, and she set to work and built a very good nest, and sat in it for six weeks, till her claws grew long and her legs grew weak, and there was of course no sign of an egg. Then I took it away from her, for I was afraid she would be ill with sitting, and it would never be the least use. Poor Thyrsis! under other circumstances she might have proved herself, if less vulgar, quite as effective as Mummy in building and breeding. When I had had her about seven or eight years she died quite suddenly. Was it of a broken heart? Had Jack's too late attentions stirred in her the emotion of love, as he clung to the corner of his cage, singing to her and leaving his babies to starve?

There is just one more canary I must mention, for it had a curious name and history. It was called after one of my relations "Uncle Arthur"; that is to say, it was called so by myself and my brothers; for it was supposed to be called "Arthur" by my mother and "Mr. Sidgwick" by the outside world.

Uncle Arthur was Jack's brother, but Jack had a monopoly of the intelligence of the family. Uncle Arthur had been half starved when he first came to me, and it had affected his intellect. Perhaps I had better mention that it was not from any supposed similarity in this respect that he was named after my uncle. He was idiotic in strange ways; for instance, I have known him try to bathe in a draught, from which he got inflammation of the lungs. For a long time, also, I found it was quite safe to take him out of doors without clipping his wings, for he was too foolish to know how to fly. One day, however, he astonished me by suddenly flying up into the top of a tree, which proved that his apparent powerlessness was the result of idiocy; for when he happened, as thus at intervals, to hit upon the right way of using his wings, he could fly quite well, though in a rather curious manner and with a pigeon-like noise. He never seemed to want to build nests, he never even serenaded any of the hen-birds of Jack's family. He had a very happy, limited life. When he was already getting old I gave him away. I am sorry to say that his death was compassed accidentally by his new mistress; she was so much disgusted with him because he would not wash [he had probably forgotten how to], that she washed him one day herself with soap and flannel. Uncle Arthur died of it.

Jack outlived all the rest. Towards the end of Mummy's life all illusion about her passed away; he got irritated and used to pull feathers out of her, though he tried to make up by much affection between times. But it was not Mummy's fault. She was frankly vulgar from the beginning, and Jack, with his keen perception of character, ought to have known it.

VII
A REGULAR FLIRT

GYPSY was so called because he was bought off a gypsy-cart. A friend of mine was attracted by his wonderful voice, and gave a half-crown for him. Others were attracted by his voice too, with results more fatal.

He was in his first year when I had him, and it was not until the second year that his feathers and his fascination attained their full proportions. Gypsy was a mule, a cross between a goldfinch and canary. His back was dark green, he had a yellowish breast with dark splashes on it, black wing feathers, and two patches on his cheeks the colour of gooseberry fool; and he had a reddish golden crest, which he could raise a little when he was excited.

The next summer was beautiful weather at Oxford, and I took Gypsy there when I went to College, though I cannot say that he aided study. If I read, he got up a quarrel with the leaves of the book, and flew at them as I turned them over. If I wrote, he fell into a passion with my pen, and ran across the wet ink on my paper to peck it. And his love-affairs were very distracting.

Gypsy's cage used to be put all day on the window-sill; and I began after a time to be aware that he was liable to be seized by sudden agitations, when he fluttered backwards and forwards in his cage, with a quick, excited note. A few days more and the cause of this agitation became apparent; for a little goldfinch, a hen goldfinch I suppose, came and sat upon the window-sill.

The intimacy rapidly improved; the goldfinch would come into the room and sit on Gypsy's cage; it made friends with a siskin and a bullfinch in the next room, and would roost in an empty cage there at night.

Gypsy's wing-feathers were clipped, so that I could let him walk about out of doors. When I took him into the garden he called to his friend, and the goldfinch dropped down by his side to take a walk with him. Other goldfinches came sometimes, but only one constantly and fearlessly when I was there. One day I remember Gypsy walking down the path in front of me accompanied by three friends.

But it was not long before there was a signal of danger. The house we were in was having some rooms added on to it, and there were workmen about. One day when I was sitting in my room and Gypsy was having an At Home, there was a little sound outside, and a limed stick was gently shoved towards my window-sill. Of course I remonstrated, and of course I was told by the workmen that they had done it entirely for my sake, because they thought that I should like to have the bird in a cage,—I could have caught the bird ten times over if I had wished it.

But this, I fear, must after all have been the end of the love-lorn bird; for it disappeared suddenly, and I never saw it again.

For a long time Gypsy had no society but mine and the canaries. He did not care for canaries, and he was mostly in a passion with me. But after some time a pair of goldfinches was given to us, much attached to each other and otherwise uninteresting. One day I put Gypsy in their cage to see what would happen. In three minutes a complete change had been worked in that happy home. Gypsy was sitting with the little lady on her perch, whispering sweet nothings into her ear, while her disconsolate spouse sat by himself on the perch below, meditating pistols for two and coffee for one.

I will do Gypsy the justice to say that he admired himself quite as much as anyone else admired him. When he was held to the looking-glass he did not fight his reflection as some animals do, he fell deeply in love with it, and whispered to it in a tiny, sweet, wooing voice, until it was obscured by a little circle of damp breath on the glass.

Some one may ask why, if Gypsy was so universally attractive and so extremely susceptible, I did not provide him with a wife to himself. Simply because it would have been no good; Gypsy was a mere flirt; he never would have had nests and eggs and brought up families like other birds; he was a mule-bird, and they cannot be domestic.

Gypsy had one last flickering of flirtation. I took his cage out one day into a London garden, and sat with him under a tree, and he sang loud; suddenly I heard a sound very unfamiliar in London, the voice of a bird which was hopping about on the tree above. I looked up, and through the leaves I could see that it was a little goldfinch; but it was shy and flew away.

These mule-birds die generally very suddenly; and Gypsy died without apparent sign of illness at about the age of ten years.

VIII
A FAITHFUL FRIEND

WE were called into my mother's room one day, and shown a hamper which had just arrived. The hamper was strangely agitated, like that hasty-pudding in which Tom Thumb sheltered, and when it was opened out rolled a puppy! It was a collie puppy, long haired, black, with tan cheeks, a white tip to his tail, white collar and paws, and wholly fascinating.

It was really a charming puppy; at present too young to sin; too young to do anything but roll about and be petted.

He was named Watch, "for," said the friend who gave him, "he is a sheep dog, and you are a pastoral family"—a very pretty reason, but I think she was also influenced by the *horlogerie* of our namesake.

Time passed, and Watch grew older and uglier. His neck lengthened, until his ears looked like ridiculous ornaments on the top of it, his legs grew long and lanky, his coat grew thin, and he grew naughty. He did not indeed eat up slippers, which is the favourite employment of story-book puppies, but he did pull most of a cold Sunday dinner on to the lawn, lick the butter out of the dish, and leave joints of mutton and beef on the grass. And he had another very original, reprehensible, natural impulse—he wished to garden. His method of gardening was to dig up saplings from a carefully-planted hedge of yews. He knew it was wrong, but he could not help it.

When he was seen thus employed, he fled back and sheltered himself in his stable. He was just in that state of mind and body which answers in human beings to the condition of rapid growth and dissatisfied temper, when sleeves retreat up the arms, and frocks and knickerbockers up the legs, and the family seems to be in a conspiracy for making things disagreeable to you.

So it seemed best that he should be sent to a shepherd for training. He went, and three months passed, and we looked daily for his return; when one morning, I was sent for to the door, where I saw, held in a strap, a beautiful, bashful, silky collie, small and well-proportioned, with long tail and ruff, and silk-fringed legs, ready to hide his face against the first friend with affection. I could hardly believe it was Watch—he was full-blown, come out!

That he should sleep in a stable any longer was a manifest impossibility. Watch was established as a house-dog.

He was wonderfully quick and obedient; he learnt to shut the door, play the piano, shake hands, catch things from his nose, and lie dead, in no time. He was so gentle that one could put little animals under his charge; the canary would stand on his head, and a kitten run between his paws. One of our blue-eyed white kittens, granddaughter of the formidable cat Rector, attached herself warmly to him.

But there were one or two circumstances under which he was not docile. Soon after he came home we took him for a walk in the fields near the town. He followed quietly; when, suddenly, he spied a flock of sheep feeding, and up went the white tufted tail like a banner; nothing could hold him; no threats restrain him, until from hedge and ditch he had collected the whole flock into marching order. Much severe treatment was necessary before we could induce him to relinquish his profession. Then often as we went through the fields, Watch following with an eager eye, longing to be off after a scattered flock, an old north-country shepherd would sidle up and "pass the time of day," and gently turn the conversation until he could say, "I suppose that dog of yours is not for sale?" He was right, Watch was not for sale.

He could not, it is true, quite resist the instinct of the chase; and often one saw him flying down the garden in pursuit of the white kitten Midge, while her old-fashioned, under-bred, good-hearted tabby mother followed to protect her. But nothing happened; he rolled over and over with Midge, and Jenny jumped upon the soft heap, and dealt out boxes of the ear when Watch's head got uppermost. Then they all got amicably up together, and went off quite good friends. Once, I am sorry to say, he did break the leg of a rabbit, but he was more surprised than any one else at it. I found him another time, having caught a blackbird; he was very much surprised and delighted, but puzzled as to the right course to adopt next; so he made short runs at it, and pretended to bite it, and wagged his tail very much, and asked me to come up and look at it.

As for the goat, he was a most excellent good comrade with her. He exercised all his sheep-driving skill to fetch her when she lagged behind. And it takes as much skill to fetch one goat as fifty sheep. When she behaved well, he consented to go in double harness with her. The double harness was made out of tape dyed purple with Judson's dyes. There was an old madman who lived in a house opposite the field where I generally drove them. He was very fond of watching the performance.

But now I come to a part of Watch's character which I cannot present in such a favourable light. He was jealous.

Of course we did not find it out at first. He was not brought into comparison with other dogs, only with inferior animals, and he would naturally not be jealous of them. We are not jealous of our friend's cat and dog, but of our friend's friends. Watch was not jealous of our cats and birds, and goats and guinea-pigs, but of our dog-acquaintance. Occasionally he showed slight uneasiness when a horse or a baby was much noticed; they were rather too high in the scale of creation—nearly at the level of dogs.

But one day there had been a dog show near us, and after the booths had been taken down, and the exhibits gone, one poor spaniel was discovered who had lost his friends, and appealed to us for sympathy; so we invited him to afternoon tea in the garden. Watch came to tea as usual; but when he saw the other dog, he suddenly became demonstratively affectionate. This was quite appreciated; but the other dog was not therefore neglected. So Watch bit him. This was not appreciated at all. We told Watch so, but he only sat down and turned his back to us, and gave the family five minutes for repentance; and as they did not fall on their knees, and beseech his forgiveness, he solemnly marched away into the house and lay in his master's study, quite alone, sulking. I am sorry to say, too, that he conceived occasionally the most violent antipathies to the most delightful and well-intentioned people. There was a friend of ours, devoted to dogs in general, and to him in particular, whom he would not allow to touch him; he would not take food from her hand; once, when he had accepted from some one else the food he had refused from her, he stopped eating it because he heard her laugh. Once he was the victim of uncontrollable fascination. A girl came to tea, at whose greeting he growled; then he lay down in a corner with his eyes fixed on her. She went on talking and taking no notice of him, and he came out into the room, little by little, looking at her, till he finally sat straight in front of her, with his eyes fixed on hers; and there he remained until she went away.

Watch had become identified with the family, to the extent of being called "Watch Benson" by many friends. His English vocabulary was wonderfully large. I remember the surprise of one gentleman who came to talk business with my father. Watch was in the room, and, hearing our voices outside, suddenly started to the door, which was shut. "Why don't you go out of the window, then?" my father said, quite quietly, and Watch in a moment ran to the window and jumped out.

I never quite knew what Watch's position was towards religious exercises. I think he approved of them, but disapproved of our exclusiveness about them. So he pretended altogether to despise church. He was depressed on Sunday morning, came to the garden gate to congratulate us when church was over, and pretended to be sleepy when the time for evening church drew near. But I think that was because he was not allowed to go; for he took up a very different position about prayers; he insisted on coming; he had his own stall in a window; though occasionally, when strangers were there, and he could not be turned out, he suddenly decided to leave it for the softer rugs in the middle of the chapel. There was one memorable occurrence, when the 26th chapter of St. Matthew was read, and Watch got more and more excited as he heard his own name repeated more and more emphatically, until at the final, "I say unto all, *Watch*," he ran eagerly out into the middle—such exciting, personal prayers!

But he made a great point of attending; for when we changed our house, and came to the conclusion that his presence would no longer be appreciated, his efforts to attend prayers were quite pathetic. Sometimes he scratched at the door, or pushed it open, and marched in in the middle; sometimes he slunk in when we went into the chapel, and sometimes ran in first and tried to hide. He had a vague idea in his mind, that it was some special privilege, some special identification with the family.

Now that we were in London half the year, Watch could not be with us constantly. For one thing his dirty paws were such a mortification to him, and we thought he would die from the amount of soot he licked off. And he could not go walks, for he would stand smiling at us in the middle of the street, with a tram, two omnibuses, a cart, and four hansoms, bearing down upon him. So he went to stay with friends, or down to the farm in the country.

That last was often necessary, but not a great success. Watch was very exclusive; he never would go walking with servants, except when everyone else was—not out, for he might have met them—but away from home. The one exception was when the servants were nurses with children. He was fond of children, and did not think it *infra dig.* to play with them. In the same way he despised everyone at the farm, and had to be treated in a very special manner, quite different from all other dogs. "Why can't Watch live like any of the other dogs?" one of the children asked. "Oh, my dear, Watch is much too good for us," his mother told him, with a deep sarcasm. No other dog could come on the rug when Watch was lying there. The cat might come and was welcome, and liked the benevolent old gentleman. Just as one would not like anybody to come and take half of one's armchair, but might be rather flattered if a cat or a little dog jumped up to settle itself there. Cats were only cats, and fit subjects for philanthropy, but other dogs were his own ill-bred relatives. As some one summed it up, "Watch doesn't care for dogs."

The other dogs could not be expected to appreciate this, and Watch's airs provoked at last one outburst from King, the steady old patriarchal collie of the farm. King flew upon him one fine day to have it out, and all the other dogs, seeing that King "had taken out a free ticket," as the bailiff phrased it, flew to avenge their private grievances. Watch was very nearly killed, but he kept his airs to the last. Such strong arguments were brought to bear upon King, that ever after, when Watch crossed the yard, King retired promptly to his kennel. He could not trust his own self-control, and fled temptation.

Poor King! he had a sad end. He and a young golden collie called Pat went out together in some woods—poachers, I fear. Towards evening Pat came back in a fearful state of agitation, trembling. The dog must have longed for words to tell what he had seen! But they guessed it. The gamekeeper was known to have a grudge against King, and he was never heard of again from that day to this.

Watch had a very different end. He grew old and blind. He had to live altogether at the farm now, but he did not mind that. He had two great friends. One was the bailiff's daughter, and one the niece of the landlady at the "Cricketers," over the way. The first nursed Watch, the second he went to see every day. But the niece got married, and Watch never crossed the road again, but transferred all his affection to Katie. He was nearly blind now, quite deaf, and very rheumatic. He had not much emotion left; it soon wearied him. I remember while he was still at the house, that when we all came home at the end of the holidays in two detachments, he greeted the first-comers effusively, and then retired under the sofa, and took no notice of the second batch until they had been in the house about an hour; then, his emotions being rested, he came out and greeted them too with affection.

But two loves remained to the end; his love for Katie and his love for milk pudding—and Katie generally gave him the milk pudding. He hobbled about after her as long as he could, and sat in her room. Once they thought him dying. He lay on Katie's bed, and Katie was away—was coming back that evening. His head lay on the pillow and his eyes were closed, and they thought him dead, when Katie came upstairs and spoke to him; and the life came back to him, and she fed him, and he lived a few days more. Then he died, this time with Katie close to him.

He is buried by the gold-fish pond under a cedar, and he has a tombstone and an epitaph, *"Esne Vigil."* And the other day I passed by, and freshly-gathered daisies were lying on it. I think Katie must have put them there.

IX
KIDS OF THE GOATS

THEY were Zoe and Marcianus Capello (but she was no kid), and Capricorn and his brother, and Chat and Tan. I did not possess them all at the same time; in fact, I never had more than three at one time, and that was because Marcianus Capello had twins.

Zoe was the first. When she came to us she was a little white kid, just taken from her mother; she was very pretty, with a dark mark down her back and two little tassels of hair on her neck. But, as I say, she was only just taken from her mother, and the first evening was full of much trouble and care, for we could not find anything she would eat, and we thought she would be starved. She would not be fed, moreover, with milk, and we were in despair until we thought of trying if she would eat the tender sprouts of may. It was early spring, and for a day or so all her meals were taken in our arms, as we held her up to nibble at the hawthorn hedge.

But she soon grew less fastidious, and, as goats do, would eat anything, from garden flowers, laurel leaves, and cabbages, down to paper and bread. She was tethered in the field, and this was very necessary, for if she was free she would follow us everywhere, would go walks with us out of doors, and would come into the house after us. The chief difficulty with kids is superabundant affection; they wail pitifully when one leaves them alone, and cannot be persuaded that their presence is not always desirable. Some friends of ours—they were Quakers too—used to dress up a stick with a waterproof and hat to keep their kid company. It satisfied her completely; but was it quite consistent with the Friends' idea of truth?

Zoe nearly had a bad accident once, in consequence of her fondness for coming into the house. I was sitting on the steps at the door and playing with her, when suddenly she bounced away from me and ran into the drawing-room. I pursued her, and she, knowing she was wrong, ran farther, saw a way of escape, and jumped straight through a large plate-glass window. I thought she would be cut to pieces, and in agonies rushed outside, where I found her making the most of her opportunities by devouring our best rose trees under the window.

Zoe lived with us for a year. Then I was kept indoors by a bad cold and throat, and I heard that the kid was sympathising with me by having a bad throat also. But alas, poor Zoe! her throat was much worse than mine, and though we strapped a little blanket on her back to keep her warm, and though the gardener and the cowman poured gruel down her throat, when she could hardly swallow, she got thinner and weaker, and one morning she died. I sobbed audibly all through prayers that morning.

Then a friend of mine told me that some cousins of hers were anxious to part with a goat they had, and Marcianus Capello, otherwise called Marcap, arrived. She was not at all like Zoe; she was a large, dull, elderly, brown and white goat. She did not want to make friends at all; she chiefly wanted to eat. But there was one great advantage about her, for a few weeks after I had her she gave birth to two little twin billygoats—two fluffy black-and-white creatures with huge legs.

Marcap was, like Zoe, tethered in the field, and it was supposed that filial affection would keep the kids near her. The kids had a fine time in consequence. One morning one heard a rattling over the roof of the nursery, and found the kids were playing King of the Castle on the house-top. Another time they came skipping out of a yard where building was going on, covered with lime to the tops of their legs; and for some little time we were terribly afraid that the smallest kid would lose his eyesight, as he had splashed lime up into his eyes, and that they both would come out of it with skinny hairless legs. A procession, of my nurse holding a cup of milk and water, myself and my youngest brother (who was too naughty to be left alone), could be seen crossing the field three times a day to bathe the kid's eyes.

When the kids were old enough to do without their mother, we gave Marcap away. I did not mind parting with Marcap; I never should have got fond of her, for she had no idea of intimacy. But to part with a kid was a different matter; it took us a long time to decide that it would be better to keep the biggest and strongest kid, Capricorn; and we gave away the little one.

Capricorn proved just a little more warlike than it is quite convenient for a kid to be, if you are in the habit of taking it out for walks. In the first place, if he met a flock of sheep in a field, he would at once begin to drive them away, running and butting after them. In the second place, if he met cows, he would invariably have a pitched battle with them, unless he was dragged away by main force. I have seen him in the middle of a ring of cows, knocked down by them, and getting up to butt them again. Thirdly, if he met a donkey, even in a cart, he would go for it, which sometimes caused the drivers of the cart to swear. Lastly, if he met children, he would try to awe them by standing on his hind-legs. His wickedness gradually developed with his growth. Before he was grown up he was a very affectionate kid. Once, when I turned back in a walk, the rest of my family had the greatest difficulty in inducing Capricorn to go with them. He got on very well with our wise collie. Watch was useful in fetching him up, if he lagged behind in a walk to carry out some of his evil designs. I had a little cart for Capricorn, too, and made him pull up stones for a rockery we were making; this was a good outlet for his energies, and he had less time to be wicked.

But he finally got too fierce for us to keep him any longer. If I was running down a hill by his side he would try to hook me with his horns, and he was not at all to be trusted with children. I gave him away reluctantly, and it was some consolation to hear that he nearly killed his new master, who came upon him suddenly in the dark. Since then I found out that it was not individual wickedness, but, so to speak, class wickedness, and that it is rarely safe to keep a billygoat when he grows up.

Then for some time I had no kid. After a while a lady near who kept goats gave me two kids.

These were very pretty kids; one was quite white, the other fawn colour, and very graceful. They would follow me everywhere; but, as I could not keep two, Chat, the white one, was given away.

It was considerably easier to take Tan walks than it had been to take Capricorn; for Tan did not want to fight every beast or child she met. Watch was useful in fetching her as he had been with Capricorn. Long afterwards, when the acquaintanceship between them was a thing of the past, to say, "Watch, fetch the kid," would bring her hurrying up to us. Tan was the only one of my goats who ever learnt a trick, but I taught her to shake hands in exchange for leaves or oats.

Then we moved from the place where we were living, and I left Tan behind me for a child of the family who were coming into our old house. I heard no more of her for a year, and then they wrote to me to say that Tan was pining, and they wished that I would send for her. So she came up by train, and the first moment she saw me she remembered me, and we shook hands.

Tan is still alive. On misty summer mornings, one sees her pass the windows heading a herd of cows; she is much too proud to walk with sheep; and though she will condescend to go with cows, she keeps herself to herself, never talks to any of them, but preserves a proud and solitary position. On rare occasions a sudden burst of friendship or curiosity will induce her to come into the house with me.

But my friendship with Tan, I must confess it, is not what it was; perhaps it might never have waned if I had not consented to the year's separation. But although occasionally we bleat to each other from a distance, though we shake hands over a few oats, she no longer runs to meet me if I come near, she no longer cries out with a wailing bleat when I go away, she no longer has to be tied up to prevent her following me. And I do not think it is age that has made this difference, I think it was worked by that year of separation.

Passing through the farmyard on a cold day, I found Tan in the corner where the dead leaves had blown up, and lay a foot or more deep. She was standing in the deepest part of the heap, which came up to the top of her legs, and had secured herself, as it were, a good hot bottle for the night.

In conclusion, I would say that there are no pets more enchanting than kids. They will give you as much amusement as kittens or puppies; while they are as intelligent as grown-up dogs, and even more wildly devoted. But there are two things you must never expect of a goat,—neither the least unselfishness in their affection, nor the smallest spark of benevolence.

X
COMMUNITY LIFE

OUR old cowman Callaway was Cornish; he taught me to milk; he took a fatherly interest in my animals; he talked Cornished English, and I understood about a quarter of what he said. He had a wife who worked in the house of a neighbour of ours, and a very elegant daughter. I never could imagine how her hats and jackets and dresses got into the hovel in which the family lived; however, I suppose they must have got into it, for they certainly came out.

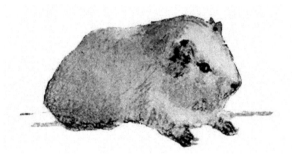

The wife's employer's daughter kept guinea-pigs; and Callaway promised to get us a white one. In due time he appeared with it. But to our delight, when the box was opened, out came two little white creatures, with shining red eyes, not weak bluish-pink eyes, but real good red ones like little jewels. They were named Ixtlilxochitl and Atahualpa, and installed in a wooden house with a wired-in yard under the laurel trees of the drying-ground. Here they rapidly became naturalised; burrowing under their wire fence, they found the way to the long, fresh grass beyond, and enjoyed as much liberty as they wished till nightfall, when the wooden slide of their house shut them safe from dogs and rats and cats.

I had many sympathisers in my amusements. Not only was there Callaway the cowman, who became house-builder to the community, but my old nurse used to take the guinea-pigs a breakfast of soaked bread every morning; and we had a butler sagacious about animals, to appeal to as a highest authority on all difficult questions. So when, one morning, I opened the slide, to find two new white things about as big as large mice gaily running about, the first thing I did was to run to the servants' hall and summon the butler to advise in this difficult and delicate situation. Ixtlilxochitl was sent to a new hutch, hastily erected for him, and Atahualpa kept house for the babies.

This was very good for the development of Ixtlilxochitl's character. He became very tame, learnt to sit up with his forepaws on my finger, and to "lie dead" on his back with his little pink hands and feet in the air; guinea-pigs' forefeet are really small pink hands, with short claws on the fingers, and a rudimentary thumb.

Guinea-pigs grow up very soon; they have no helpless infancy at all. I have heard of a guinea-pig eating bran twenty minutes after it was born. I know we used to carry the infants about and let them run up our sleeves till they stuck, and had to be pulled back by their hind-legs; and though I would not recommend this practice, they never seemed to take any harm from it. Then, when they are about three months old they become heads of families. At first the family only consists of one or two members, but they increase in number until each family numbers seven or eight. You may expect a new family once every six or eight weeks. There is a nice sum in geometrical progression! And after this general statement of the matter you will hardly expect me to give you a history of each individual, though I made a chart of their genealogies. I will, however, give a short biographical notice of the most interesting characters.

The first two were Ulfias and Brastias. Ulfias was a nice, comely guinea-pig; he took after his father, and had brown whiskers. Brastias had pink ears, which were generally much bitten, and fierce red eyes; he was an ill-conditioned, cross little beast, and a great fighter. Moreover, he was a murderer.

It is the funniest thing in the world to see guinea-pigs fighting. They stand on the tips of their toes and raise their noses, until they present the chin only to their adversaries; then they begin to dance round, always chin to chin, gnashing their teeth; when they see a good opportunity they fly in and bite. It is a scientific way of fighting, like wrestling or fencing—quite different from the indiscriminate plunge of a cat, who rolls round in a heap with her adversary.

After these two came Enid, Elaine, and Geraint. Enid was the first to have a baby, and she had only one—a fat round one, which grew and prospered until one day when he suddenly disappeared. We searched and hunted with anxious hearts, but with no result. After a time we wanted to move the hutches to a new place, and when we took up that in which poor Enid and the baby lived, there was a hole under it—a rat's hole, and at the end of the hole, as we peered down, we saw a little white thing—the skin and bones of a baby guinea-pig. Enid never had another baby; she grew sad and thin and pined away, and at last she died.

Then Elaine had a baby—two; but one was deformed, completely paralysed in his hind-legs, and I felt that the kindest thing to do would be to destroy him. So I took out a bottle of laudanum, and prepared to begin the hari-kari. Poor little guinea-pig! it was already very ill, and I could with difficulty get its little rabbit-like mouth open. What a tiny throat! could it swallow even enough poison to end its panting little life? When I laid it down again there was very little change, and I did not know what to do; then the pink nose, the hands and feet, began to have a slightly blue tinge. I could not disturb it again to open its mouth, so I poured a little more laudanum on its mouth and nose, and the limbs got bluer, and the breathing became harder, and at last ceased. It was a dreadful thing to do. However, on the whole, it was less dreadful than drowning it. Once I had to drown a bat.... We will draw a veil over that.

However, to proceed with the guinea-pigs. The baby that was not deformed was a very nice little pig—small but comely. He grew up and was called Jim.

There is an individuality about guinea-pigs, not explicable but to be apprehended intuitively. Jim was quite individual. You would have known that if you had only seen him sitting upright at his mother's side to nibble out of the hay trough.

The guinea-pigs lived in a large estate fenced in by wire; inside the yard were various settlements, bedrooms, all with free access to the yard, and usually to the ground beyond, for they made holes under the wire and disported themselves outside. They had a beautiful rack to hold their hay, saucers for bran, and were given a breakfast of soaked bread every morning. At breakfast-time shrill whistles might be heard from the guinea-pig yard. Most people think guinea-pigs have only one noise, but in reality they have, quite clearly defined, three fundamental notes, of desire, contentment, and anger. They whistle when they are hungry, make what are called "guinea-pig noises" when they are well content—for ordinary conversation, and they gnash their teeth when they are angry.

About this time, when the colony was not too large, I used to take them out for picnics.

Opposite the front door, at the corner of the lawn, there is a large escalonia tree; on warm summer evenings it sheds a delicious fragrant smell from leaves and flowers. Opposite this there is a stile made to get into the fields. The stile is made in such a manner as to be a very comfortable seat. Here, under the escalonia, I used to turn out the guinea-pigs for a day in the country, while I read a book on the stile, and Watch was put to guard them; if any little pig strayed too far, he saw where it went to, and helped me to find it again.

But, in time, the colony grew too large for this, and at last it began to increase with a rapidity that alarmed me; for, as you see, it is not a case of the simple geometrical progression of creatures which have the same number in every family; but, as guinea-pigs get older, each family gets larger, so that it is like a sum in compound interest, at an accelerating rate of interest. I began to be frightened when the "five Mitchinsons" were born, and the next family was larger still.

In fact, they would have eaten us out of garden and farm, if it had not been for what political economists call "violent checks"; these violent checks were kidnapping, nepoticide, and massacre.

Kidnapping was the first check. Our house was being added to, and there were various workmen about, and one morning when I visited the hutches, Daisy and Ally Mitchinson were missing. There is no more to say about it; they were never seen again. I felt like a mother, who, having complained of the burden and size of her family, is deprived of one of them.

But that was not the worst. Atahualpa was still flourishing, although a great-great-grandmother. One morning I found reason to seclude her from the rest of the community, and by an arrangement of hutches, I shut off a little yard for her by herself.

I came back a few hours later, and I found Brastias had displayed himself in his true colours at last. He had leaped the barrier, and was standing with gory mouth and fiery eye, over the carcase of a baby guinea-pig. In another corner of the hutch was Atahualpa, behaving with the supremest indifference to six more.

That day I gave away sixteen guinea-pigs. But I believe that we should have had a repetition of Bishop Hatto, if it had not been for the last check—namely, massacre.

We were overrun with rats, and rat-catchers were sent for. One morning two men came up with their dogs. The men were looking at the rat-holes, and arranging a plan of campaign, when suddenly they found that the dogs were not with them. Across the wall which separated the cow stables and haystacks from the garden and guinea-pig yard, they heard a doleful noise. They ran round, and found that the dogs had been doing their duty nobly, and all the guinea-pigs but two lay dead on the ground.

The victims were buried in a large grave, and my brother found a suitable slate and wrote a Latin epitaph on it. He put it up as a headstone, and enjoyed the proceeding very much.

But I did not enjoy it. I had not the heart to keep guinea-pigs any more. I gave away the two survivors, and the hutches mouldered away, and cucumbers grew over the yard, and only the genealogy and the tombstone were left as memorials of that very large family with the white coats and jewelled eyes.

XI
FINISHED SOLOMON

KING Solomon was journeying through a thirsty land—sand beneath his feet, sand around as far as a man could see, above the pitiless blue sky. No tree could grow here, and no rock was there to cast its shadow on the sand. "What shall shield me," said the king, "from the fury of this sun?" Then was heard the sound of light wings beating the air, for all creatures knew the voice of the words of King Solomon; and there came through the air a cloud of hoopoes, and they spread their barred wings, and closed them together, wing to wing, and they shielded King Solomon. So, when the toilsome journey was over, the king called the hoopoes, and said, "O hoopoes, what will ye that I give you for your service done to me this day?" And the hoopoes said, "O King, give us crowns of gold"; and the king gave the hoopoes crowns of pure gold.

But men hunted the hoopoes through the length and breadth of the land, and they killed them for the sake of their golden crowns; then the hoopoes cried to King Solomon, for King Solomon knew the voice of all beasts and birds, yea of the creeping things also, and the hoopoes said, "Take away our crowns, O king, for men kill us for the sake of our golden crowns." And Solomon took away their crowns. "Yet," said he, "it shall be known what the hoopoes did for the King," and he gave them crowns of golden feathers.

So says the *Book of the Enchantments of the King*, and that is why my hoopoe was called Solomon.

I was riding through a village near Thebes in the evening, and among the groups of children who held out grimy hands and cried "Backsheesh"; and the half-blind boys who made the somewhat startling statement, "Finished Fazzer, finished Muzzer, I yam berry hongerry"; I saw at the door of a mud house three children, one of whom swung towards me a bird he held by the wings,—and I recognised the helpless, half-dead, fluffy mass for a hoopoe.

I refused to give them the wages of sin, and they were too much surprised to attempt to hinder the departure of the hoopoe. Indeed, if they had kept it much longer, it would have departed without assistance by the silent road, for one claw had been tied back to its leg, and it had been swung in that manner till its tormentors happened to think that they had better try the wings instead; its crown of feathers had been pulled out; and when I got back to the hotel, it shut its eyes and fell forwards on the point of its beak as if it was about to die. The string had been tied so tightly that it was with difficulty that we got it free from its bonds, and then we plied it with whisky and water. That was no easy matter either, for it would not open its mouth, and one had first to get the long beak open, and then to hold it so, while from a feather dipped in the refreshing beverage a drop trickled down the pink throat; then the bill was shut, and one watched to see if the feathers of the throat would ruffle and give sign that the drop was passing down. The method succeeded, for presently the little forked tongue was shot out to suck up the liquid, the little brown eyes opened, and the hoopoe, taking in the situation, hurried into the corner of the window-sill, and supposed that he was hiding himself by laying his long bill up the wall.

It would certainly be necessary to provide the hoopoe with a habitation, were he only the guest of a day; so a crate which had contained pottery was found, its straw was arranged nestwise, and the bird was bestowed in it, much to its own satisfaction.

But the diet was a problem. Its natural food was live insects. I went so far as to kill a housefly, but it was a very disgusting process, and the fly was not at all well received; moreover, I was not sure whether the hoopoe was of an age to receive, shall I say *peptonised* food from his parents, or whether he preferred the raw material. But as the best compromise, including the carnivorous and the more-or-less-peptonised element, I decided on hard-boiled egg; that had to be administered in the same way as the whisky, with drops of water to help it to run down. After this I put the hoopoe into the crate for the night.

I frankly confess that I expected to find a stiff little body there in the morning, but instead I saw a bright brown eye fixed upon me, and a smooth, compact, though crownless little hoopoe, sitting in the straw.

If the hoopoe was going to live, other things became necessary—first and foremost, a name.

The name suited him exactly. From the time that he was called Solomon, he *became* Solomon. We never spoke of him as the hoopoe; indeed, it is with great difficulty that I have avoided so far using his name. Now I have told you when and why he was named; henceforth, then, he is Solomon.

But, secondly, Solomon must have exercise, and fresh animal food. It would be better, both for the sake of digestion and economy of time, if the two could be combined, and I spent most of my time in effecting the combination in one of the garden beds.

The beds in the hotel garden are excellently convenient for feeding and exercising half-fledged hoopoes; they are lowered three or four inches below the level of the paths, for the purposes of irrigation. Thus when, once a week, the water is turned in, the beds become a series of pools, until the water has gradually soaked away through the rich black mud. Further, the beds are surrounded with a bushy little plant, so that when Solomon tried to spring over the edge and escape me, his wings were not strong enough for the purpose; he sprawled on the bushy plant, wings spread and legs kicking, and was easily captured.

But it was Sunday, and the hour drew towards church time. Solomon must go home and be fed before I went to church. Accordingly, I went to catch him, but there was one thing I had forgotten. At the corner of the bed was a drain through which the irrigation was effected. Quick as thought Solomon ran in there, and was out of arm's length in a minute. What was to be done? The bell was already ringing to church; decent and godly people, with their prayer-books in their hands, were walking down the garden path; and there was I plunging round the drain in search of an ungrateful, half-fledged, discrowned hoopoe. I dared not leave him there, to be the prey of the numerous and ravenous hawks and crows.

But suddenly, as a *Deus ex machina*, Mahmoud the gardener hove in sight; so I called to Mahmoud, and Mahmoud called to Ibrahim, and Ibrahim brought a dry palm leaf, and we put it in at the opposite end of the drain, and made a very terrible shaking noise in the inside with it; and there hurried out a very long beak, supported by a very small bird at the end of it; and Solomon was captured in time for church.

When I came back from church, Solomon's crate was empty. We trod carefully over the room for fear of squashing him flat, like a botanical specimen; we looked under the sofa, under the chairs, and Solomon was not there. Then a little scuffling noise on the balcony attracted our attention, and there was Solomon with a guilty look in his face. We lined the inside of his crate with stiff newspaper.

But when I came back from lunch I saw a ridiculous silhouette far up the half-lighted passage. There again was Solomon! He had carried on mining operations on the paper during lunch, and had escaped again. Another crate with narrower bars had to be procured. Of course he instantly put his head through and got it fixed, and I had to seize him by the beak and push him back.

Now, by all the laws of animal literature, Solomon ought to have been devoted to me by this time. If he had studied the *Whole Duty of Birds*, he would have found out that he must wake me at dawn (I cannot feel sure that I should have appreciated that); that he must flutter his wings with joy and chirp when I came into the room, even if he did not feel equal to opening his little bill and pouring forth a grateful song (do hoopoes sing?); that he must follow me round the room; that he must eat out of my hand; that he must beat his breast against the bars of his cage when I went away.

Solomon did none of these things. He shut his beak tightly when I wished to feed him, he pecked at me when I tried to open it, he ran away when I attempted to catch him, he struggled when I had got him, he hurled himself from my hand into the crate as soon as possible, and he did not like me at all.

By the third day Solomon had immensely developed. People who had considerately told me that it was impossible to rear a hoopoe, now foretold that he would live. He extended his mining operations to the garden. I am not sure that he found any insects, but he did great execution on the loose earth at the foot of the palm-tree. He looked quite like a real grown-up hoopoe when he ran about the garden bed and dug his bill in up to its roots; and in the evening he flopped off the window-sill while I was feeding him, and had a grand race round the room.

That night I dismissed the fear of finding the little cold corpse in the morning.

But when I opened the shutters and looked at Solomon in the morning, he was not awake; his head was tucked behind his wing. I took him out, he looked round dreamily, and sank on to the ground. I got whisky and water again, and fed him with a feather; he pecked and struggled at first, but presently he allowed me to open his beak, and I saw that the little pink mouth was getting very white. Still I gave him more, hoping it would have the same reviving effect as at first. But presently Solomon dropped his beak on the window-sill, and the drop trickled down it again, for he had stopped swallowing. He laid his head down, and stretched out his little black claws; and heaved gently once or twice; and no more.

As the Arabs say, it was "Finished Solomon."

www.ingramcontent.com/pod-product-compliance
Ingram Content Group UK Ltd.
Pitfield, Milton Keynes, MK11 3LW, UK
UKHW042151281224
453045UK00004B/335